EARTH'S ROCKS IN REVIEW

THE ROCK CYCLE

By Anna McDougal

Enslow PUBLISHING

Please visit our website, www.enslow.com. For a free color catalog of all our high-quality books, call toll free 1-800-398-2504 or fax 1-877-980-4454.

Cataloging-in-Publication Data
Names: McDougal, Anna.
Title: The rock cycle / Anna McDougal.
Description: New York : Enslow Publishing, 2024. | Series: Earth's rocks in review | Includes glossary and index.
Identifiers: ISBN 9781978538009 (pbk.) | ISBN 9781978538016 (library bound) | ISBN 9781978538023 (ebook)
Subjects: LCSH: Petrology–Juvenile literature. | Geochemical cycles–Juvenile literature.
Classification: LCC QE432.2 M425 2024 | DDC 552–dc23

Published in 2024 by
Enslow Publishing
2544 Clinton Street
Buffalo, NY 14224

Copyright © 2024 Enslow Publishing

Portions of this work were originally authored by Frances Nagle and published as *What Is the Rock Cycle?* All new material in this edition authored by Anna McDougal.

Designer: Claire Wrazin
Editor: Caitie McAneney

Photo credits: Cover, p. 1 Mathias Berlin/Shutterstock.com; series art (title & heading background shape) cddesign.co/Shutterstock.com; series art (dark stone background) Somchai kong/Shutterstock.com; series art (white stone header background) Madredus/Shutterstock.com; series art (light stone background) hlinjue/Shutterstock.com; series art (learn more stone background) MaraZe/Shutterstock.com; p. 4 File:James hutton-1726-1797.gif/Wikimedia Commons; p. 5 TenchLi/Shutterstock.com; pp. 4, 23 arrows Elina Li/Shutterstock.com; p. 7 (top) OlenaPalaguta/Shutterstock.cpm, (bottom) Honourr/Shutterstock.com; p. 9 (top) DariaGa/Shutterstock.com, (bottom) VectorMine/Shutterstock.com; p. 11 (top) Ammatar/Shutterstock.com, (bottom) kubais/Shutterstock.com; p. 13 (top) Tom Gowanlock/Shutterstock.com, (bottom) trekandshoot/Shutterstock.com; p. 15 PHONGPHAT PRASONG/Shutterstock.com; p. 17 Amit kg/Shutterstock.com; p. 19 (top) michal812/Shutterstock.com, (bottom) Artography/Shutterstock.com; p. 21 DanielFreyr/Shutterstock.com; p. 23 William T Smith/Shutterstock.com; p. 25 Ursula Perreten/Shutterstock.com; p. 27 milicenta/Shutterstock.com; p. 29 Alizada Studios/Shutterstock.com.

All rights reserved. No part of this book may be reproduced in any form without permission in writing from the publisher, except by a reviewer.

Printed in the United States of America

Some of the images in this book illustrate individuals who are models. The depictions do not imply actual situations or events.

CPSIA compliance information: Batch #CWENS24: For further information, contact Enslow Publishing at 1-800-398-2504.

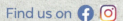

CONTENTS

An Exciting Cycle 4
Earth's Layers 6
A Look at the Lithosphere 8
What's the Weather? 10
Carried Away 12
Super Sedimentary Rocks 14
Under Pressure 16
Making Metamorphic Rocks 18
Hot Lava! 20
Inside Igneous Rocks 22
An Uplifting Step 24
Keep Rocking 26
Learning from Rocks 28
Exploring the Rock Cycle! 30
Glossary 31
For More Information 32
Index 32

Words in the glossary appear in **bold** the first time they are used in the text.

AN EXCITING CYCLE

The earth you stand on is made of rocks. But how do rocks form? In the 1700s, a scientist named James Hutton came up with a **model** that is today called the rock **cycle**. It explained how rocks on Earth form, break down, and reform.

James Hutton (1726–1797)

LEARN MORE

Rocks are matter on Earth made from **minerals**. They form naturally, and the rock cycle explains how that happens.

EARTH'S LAYERS

The part of Earth that people stand on is called the crust. It's only a small part of all Earth's rock! The other **layers** are the inner core, outer core, and mantle. The crust and the upper part of the mantle are called the lithosphere.

LEARN MORE

The lithosphere is where the rock cycle mostly happens.

Earth's crust

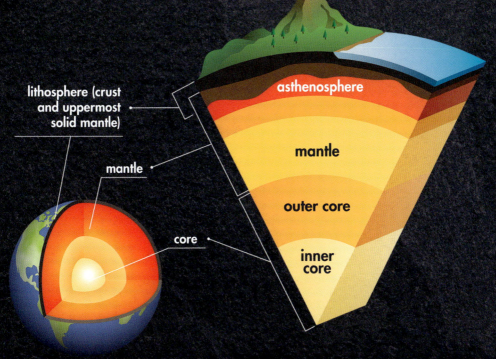

lithosphere (crust and uppermost solid mantle)

asthenosphere

mantle

outer core

inner core

core

A LOOK AT THE LITHOSPHERE

Large slabs of rock called tectonic plates make up the lithosphere. These plates move! They slide past one another, run into each other, and slip under and over each other. These movements cause events of the rock cycle to occur.

LEARN MORE

Tectonic plate boundaries, or areas where tectonic plates meet, often have landforms like mountains.

mountains

TECTONIC PLATE BOUNDARIES

divergent plate boundary

convergent plate boundary

transform plate boundary

- oceanic ridge
- oceanic trench
- transform fault

OCEAN CRUST

MANTLE

WHAT'S THE WEATHER?

The first step in the rock cycle is weathering. That's the breaking down of rock. **Chemical** weathering is when the rock's **molecular** makeup changes. For example, water mixed with chemicals may flow into rock, causing it to get softer and break down.

LEARN MORE

Physical weathering includes wind and water breaking down rock. Biological weathering is when animals cause weathering by digging at soft rock.

CARRIED AWAY

When rock breaks down, it becomes little bits called sediment. Sediment includes sand, soil, and small stones. Erosion is the movement of sediment. Sediment is often carried by wind or water. For example, some rocks are carried to beaches by water.

LEARN MORE

Gravity can be part of erosion, such as pulling loose rock down a mountainside during a rockslide.

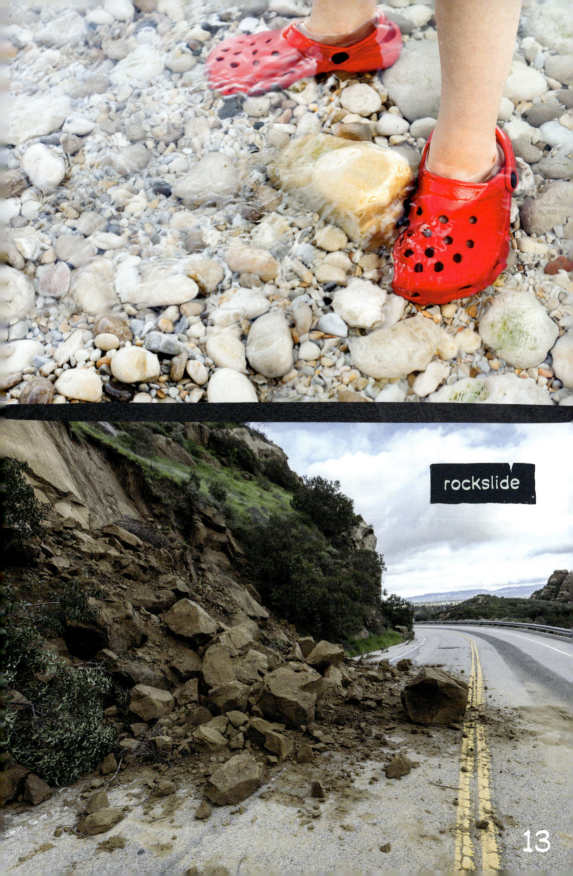

rockslide

SUPER SEDIMENTARY ROCKS

Finally, the sediment settles in a place, which is called deposition. Over time, other matter falls on top and buries it. The **pressure** on the buried sediment builds up. Sedimentary rock forms! It also forms when water flows into open spaces and leaves behind minerals that bind sediment together.

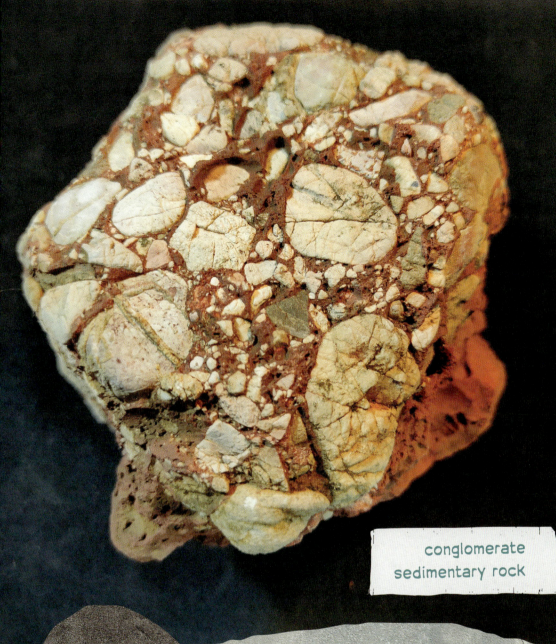

conglomerate sedimentary rock

LEARN MORE

Bits of different kinds of rocks sometimes come together in sedimentary rocks. You can see the different kinds in conglomerate rocks.

UNDER PRESSURE

When tectonic plates move, they may force surface rock down underground. Over time, this rock faces great heat and pressure. Then, the rock—called the "parent rock"—changes when its minerals become **unstable**. That makes metamorphic rock!

metamorphic rock

LEARN MORE

The deeper a rock goes inside Earth, the more heat and pressure it faces.

MAKING METAMORPHIC ROCKS

Any rock—igneous, sedimentary, or even metamorphic—can become a new metamorphic rock. While becoming metamorphic, the chemical makeup of a rock changes because of new **conditions**. The minerals in the parent rock play a big part in what kind of metamorphic rock will form

granite (igneous rock)

gneiss (metamorphic rock)

LEARN MORE

Heat and pressure can turn the igneous rock granite into the metamorphic rock gneiss.

HOT LAVA!

The deeper rock goes underground, the hotter it gets. Sometimes it even melts, becoming liquid. Hot, liquid rock underground is called magma. When this hot, liquid rock rises back to Earth's surface, it's lava. When lava cools, it forms igneous rock.

volcano

LEARN MORE

Lava comes out of openings in Earth's surface called volcanoes.

INSIDE IGNEOUS ROCKS

There are two **processes** for forming igneous rock. Extrusive igneous rock forms when lava reaches Earth's surface and cools. Intrusive igneous rock forms underground as magma flows into pockets and tunnels below Earth's surface.

LEARN MORE

Extrusive igneous rocks are also called volcanic rocks. One example is obsidian.

23

AN UPLIFTING STEP

Have you ever climbed a mountain? Mountains are formed through uplift, which is when tectonic plate movement raises one part of Earth's surface over another part. This creates mountains, where rock faces weathering again. Then the whole rock cycle can start over!

LEARN MORE

The tallest mountain in the world is Mount Everest. It was formed around 60 million years ago when two tectonic plates came together.

Mount Everest, Nepal

KEEP ROCKING

The rock cycle has no real beginning or end. It's been happening since Earth formed, and it's going on right now! Any of the three kinds of rock can be made into another kind of rock with enough time and the right conditions.

LEARN MORE

Geologists are scientists who study Earth, including its many rocks. They can tell if a rock is igneous, sedimentary, or metamorphic by how it looks.

LEARNING FROM ROCKS

Rocks tell us a lot about Earth! By looking at when a rock formed, geologists can learn what Earth was like long ago. They also look for fossils, or marks and remains found in sedimentary rocks left behind by living things.

LEARN MORE

Rocks tell us that Earth formed more than 4.5 billion years ago!

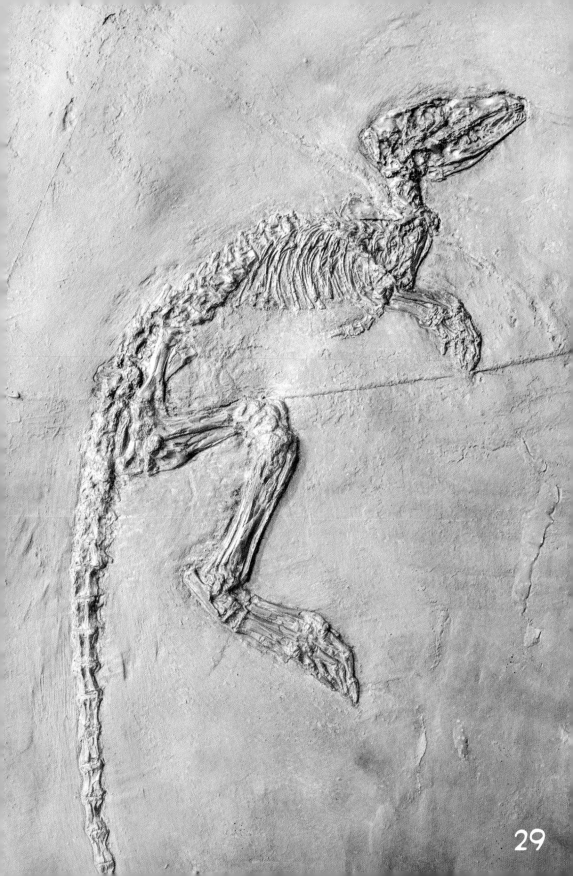

EXPLORING THE ROCK CYCLE!

surface rocks break down due to weathering and erosion

deposited underwater

SEDIMENT

IGNEOUS ROCK

compaction and cementation

SEDIMENTARY ROCK

cooling and crystalization

buried underground in heat and pressure

METAMORPHIC ROCK

melting

MAGMA

chemical: Matter, such as an element or compound, that can be mixed with other matter to create changes. Also, having to do with matter that can be mixed with other matter to cause changes.

condition: The way things are at a time or in a place.

cycle: A sequence of events that repeats.

gravity: The force that pulls objects toward Earth's center.

layer: One thickness of something lying over or under another.

mineral: Matter in the ground that forms rocks.

model: A set of ideas or numbers that explain something's past, present, or future state.

molecular: Having to do with molecules, which are very small pieces of matter.

physical: Having to do with matter that can be touched or seen.

pressure: A force that pushes on something else.

process: A set of steps.

unstable: Not likely to change suddenly or greatly.

FOR MORE INFORMATION

BOOKS

Owen, Ruth. *The Rock Cycle*. Minneapolis, MN: Bearport Publishing, 2022.

Rogers, Marie. *Exploring the Rock Cycle*. New York, NY: PowerKids Press, 2022.

WEBSITE

Geology 101

kids.nationalgeographic.com/science/article/geology-101

Learn more about rocks and what they teach us with National Geographic Kids.

Publisher's note to educators and parents: Our editors have carefully reviewed this website to ensure it is suitable for students. Many websites change frequently, however, and we cannot guarantee that a site's future contents will continue to meet our high standards of quality and educational value. Be advised that students should be closely supervised whenever they access the internet.

INDEX

erosion, 12, 30

fossil, 28

Hutton, James, 4

igneous rock, 18, 19, 20, 22, 27, 30

lithosphere, 6, 7, 8

metamorphic rock, 16, 18, 27, 30

minerals, 5, 16

sediment, 12, 14, 30

sedimentary rock, 14, 15, 18, 27, 30

tectonic plates, 8, 9, 16, 24

uplift, 24

weathering, 10, 24, 30